Math Secrets for the SAT® and ACT®

by

Richard F. Corn

ISBN 978-0-615-25399-2

ISBN: 978-0-615-25399-2

Comments and suggestions for improving this book are encouraged and welcome. Please send them to mathfeedback@yahoo.com

Errata for this book may be found at http://www.mathtutorct.com/errata.html.

TABLE OF CONTENTS

Introduction

In my tutoring practice, preparation for the standardized tests begins by doing a thorough math review using my other book, Math Study Guide for the SAT®, ACT® and SAT® Subject Tests.

Math review is followed by lots of practice tests. As we work through the practice tests, I build a custom list of techniques the student should consider using to help raise his or her score. Having made many lists for many students, I put together this book containing the techniques that have proven to be most helpful most often.

Techniques are illustrated using examples from the practice tests published by the test makers. Specifically, SAT® students use the tests published by the College Board in The Official SAT Study Guide, and ACT® students use the tests published by ACT, Inc. in The Real ACT Prep Guide.

If used as directed, these techniques should enable you to raise your score without spending hours on improving your math. They are called "secrets," but the only thing secret about them is that students do not know them (some so-called experts do not seem to know them either). They are the accumulated wisdom of having tutored high school students for the math sections of the SAT® and ACT® for several years. I track my students' scores, and these techniques have served them well.

This book is divided into two parts, one for each test. Each part is independent of the other and parts can be used in any order, either SAT® first or ACT® first. You may notice that some techniques are common to both tests, but they appear in different orders. That is because techniques are ordered according to usefulness. A given technique may be more useful on one test than the other test.

Introduction

Each part of the book has the same structure:

- The first two techniques apply to all of the problems on the test.
- These are followed by a series of techniques in the same format. First, the technique is described and illustrated by an example. Further examples of each technique are taken from practice problems published by the test makers.
- At the end of each part you will find a sheet that can be used to track your progress on the practice tests. You should be doing practice tests as you learn the techniques. With each practice test, you should become better and better able to apply these techniques, thus improving your test scores.

As stated earlier, techniques are listed in order of usefulness, with the most useful listed first. If you do nothing else to prepare, try to learn the first five or so techniques before taking the test, and take as many practice tests as you can.

Good luck on your test!

Notes

SAT® Secret #1: Solve-Guess-Skip

This technique applies to every problem and every type of student, so pay close attention.

Solve

The best way to answer a problem is through a "direct solution," where you read the problem carefully and calculate your answer using the math learned in school. If the problem is multiple choice, match your answer to one of the answer choices given and bubble that in. If the problem is a grid-in (student supplied response), carefully bubble in your answer.

Guess

If you cannot solve the problem directly, it may be advisable to guess.

If the problem is multiple choice, try to eliminate at least two of the answer choices. If you can eliminate at least two answer choices with confidence, then randomly guess one of the remaining answer choices.

If the problem is a grid-in (student supplied response), then bubble in any reasonable answer. For grid-in problems, a wrong answer is scored the same as a skip, so you have nothing to lose by guessing.

Skip

If you cannot calculate a direct solution or guess, then leave your answer sheet blank for this problem. Mark your question book so that if there is time remaining you can go back and try the problem again, especially if the problem appears early in the section.

Take care to bubble in your next answer below the space for the current problem.

More Information on SAT® Secret #1: Solve-Guess-Skip

Almost every student has trouble skipping problems. Years of experience in school have taught you that you should try every problem on the test because you may at least get partial credit. Also, if you skip the more difficult problems toward the end of the test you cannot possibly expect a good grade. You must unlearn this!

I repeat: you must unlearn this!

The SAT® penalizes wrong answers to the multiple choice problems. On tests in school, there is no penalty for a wrong answer. The SAT® has no partial credit. Just these two differences change everything, but there are other differences as well.

Skipping is the best tool you have to cope with the SAT®. It enables you to spend most of your time on the problems you are most likely to answer correctly. It can be the "secret" to raising your score.

I repeat: skipping can be the best way to raise your score.

If your goal is a 550, you could skip 23 out of the 54 math problems (43% of them!) and still reach your goal. If you tried skipping that many on a regular math test, you would fail for sure. If your goal is a 650, you could skip 11 out of the 54 math problems (20% of them!) and still reach your goal. On a regular math test the best you could get would be a B. If your goal is a 750, you could skip one or two problems and still reach your goal.

These skip numbers are just general guidelines. With practice, you will find the right number of problems you should attempt, and you will become better able to spot the ones you should skip.

SAT® Secret #2: Optimize Your Time

The general principle of this technique is to spend more time on questions that you are more likely to be able to solve and less time on questions that you are less likely to solve. It is self-evident that this is a solid idea, so why do so few students do it?

Time optimization goes hand-in-hand with the solve-guess-skip technique. If solve-guess-skip is applied consistently, then your time should be spent reasonably well. To optimize your time, shorten the test as described in more detail on the next page. But first ...more information about the test and time management.

A key difference between the SAT® and tests in school is that every question counts the same. The harder questions that appear toward the end of a section count the same as the easier questions that appear at the beginning of a section! Spend the bulk of your time on questions you are most likely to solve because there is no advantage to attempting the more difficult questions.

If your goal is a very high score (700+), you should read every problem on the test and follow the solve-guess-skip rule carefully. For the rest of us mere mortals, it is a good idea to shorten the test. This is a bit like choking up on a baseball bat. Read the next page to learn how to do it.

More Information on
SAT® Secret #2: Optimize Your Time

Within each math section, the questions on the test are ordered from easy to medium to hard, yet they are weighted equally. So the idea is to spend as much time as it takes on the easier problems to get them right, avoiding silly mistakes. The remaining time should be spent on the more difficult problems, making sure to apply the answer-guess-skip rule as if your life depended on it!

If your goal is 700+ you should read but not necessarily answer every problem. If your goal is 650, don't be worried if you run out of time with two or three problems remaining. I know that this is counter-intuitive but please try it anyway. If your goal is 550 or less, try reading only the first half or so of the section. You do not even have to look at the other problems to reach your goal! With practice, you will find the amount of skipping that works best for you. For most students, answering fewer problems will raise your score! Remember: you have to unlearn ten years of test taking.

Unless the test contains an experimental math section, there will be three math sections. Two math sections will consist of only multiple choice problems and these will run from easy to hard. A third section, called the "mixed section," will begin with multiple choice problems followed by grid-in problems. Special attention must be paid to the mixed section.

In the mixed section the multiple choice problems run from easy to hard, and then the grid-in problems run from easy to hard. Try to answer the first half or so of the multiple choice problems and then skip to answer the first half or so of the grid-in problems. Nail the easier problems in both parts of the mixed section first. If there is time remaining, you can continue with the grid-in problems or go back to the multiple choice problems in that section. You can move back and forth as much as you want as long as you remain in the section.

SAT® Secret #3: Backsolving

We now begin techniques with specific examples. The technique of backsolving applies only to the multiple choice problems. They constitute the bulk of the test -- 44 out of the 54 math problems are multiple choice.

Many students prefer multiple choice problems because the answer is right there. It is among the five answer choices and "all you have to do" is identify it.

As stated before, the best approach is the direct solution. Read the problem carefully, calculate the solution, and match your answer with an answer choice. If you are not able to calculate the solution, then backsolving is an option.

To backsolve a problem, simply plug each of the answer choices into the original problem and see which one works. Many times the answer choices are ordered from smallest to largest (or largest to smallest). If that is the case, try the middle answer choice first. If it is not correct, go up or down from there.

Some problems are structured so that they cannot be backsolved, so this techniques does not always apply. Also, backsolving takes time.

EXAMPLE PROBLEM:

What value of x satisfies the equation $2|x-3|=8$?

(A) -7
(B) -1
(C) 0
(D) 1
(E) 2

SOLVE BY BACKSOLVING:

Plug in answer choices.

(A) $2|-7-3|=2\cdot 10=20$

(B) $2|-1-3|=2\cdot 4=8$

(C) $2|0-3|=2\cdot 3=6$

(D) $2|1-3|=2\cdot 2=4$

(E) $2|2-3|=2\cdot 1=2$

The answer is (B).

Official Problems that Use SAT® Secret #3: Backsolving

page 397, problem 9

(A) $2^{2\cdot 2} = 2^4 = 16$

$8^{2-1} = 8^1 = 8$

(B) $2^{2\cdot 3} = 2^6 = 64$

$8^{3-1} = 8^2 = 64$

(C) $2^{2\cdot 4} = 2^8 = 256$

$8^{4-1} = 8^3 = 512$

(D) $2^{2\cdot 5} = 2^{10} = 1024$

$8^{5-1} = 8^4 = 4096$

(E) $2^{2\cdot 6} = 2^{12} = 4096$

$8^{6-1} = 8^5 = 32,768$

page 459, problem 2

(A) $2^{4\cdot 1} = 2^4 = 16$

(B) $2^{4\cdot 2} = 2^8 = 256$

(C) $2^{4\cdot 4} = 2^{16} = 65,536$

(D) $2^{4\cdot 8} = 2^{32} = 4,294,967,296$

(E) $2^{4\cdot 12} = 2^{48} = 2.8 \text{ x } 10^{14}$

page 471, problem 1

(A) $\frac{3+4}{2} = \frac{7}{2} = 3.5$

(B) $\frac{3+5}{2} = \frac{8}{2} = 4$

(C) $\frac{3+9}{2} = \frac{12}{2} = 6$

(D) $\frac{3+12}{2} = \frac{15}{2} = 7.5$

(E) $\frac{3+15}{2} = \frac{18}{2} = 9$

page 520, problem 10

(A) $3(-3)^2 < (3\cdot(-3))^2$

$27 < 81$ true

(B) $3\cdot 0^2 < (3\cdot 0)^2$

$0 < 0$ false

(C) $3\left(\frac{1}{3}\right)^2 < \left(3\left(\frac{1}{3}\right)\right)^2$

$\frac{3}{9} < 1$ true

(D) $3\cdot 1^2 < (3\cdot 1)^2$

$3 < 9$ true

(E) We already found that (B) is false.

SAT® Secret #4: Substitution

Substitution may be used on multiple choice or grid-in (student supplied response) problems. It can be widely used, especially instead of using algebra. In fact, unless you are an A-student in math, I strongly suggest solving problems directly using substitution whenever possible. The A-students can go ahead and use algebra. A good example appears below.

EXAMPLE PROBLEM:

A store clerk was asked to markup the price of a pair of shoes by 20%. Instead the clerk marked the price down by 20%. By what percentage does the price now have to be increased in order to be correct?

(A) 20%

(B) 30%

(C) 33%

(D) 40%

(E) 50%

SOLVE BY SUBSTITUTION:

Assume the original price was $100. It was erroneously marked down to $80 when it should have been marked up to $120. Therefore the wrong $80 price must be increased by $40 to obtain the correct price of $120. This is a 50% increase over the wrong price. The answer is (E).

SOLVE BY ALGEBRA:

Assume the original price was x dollars. It was erroneously marked down to .8x dollars when it should have been marked up to 1.2x dollars. Therefore it must be increased by 1.2x -.8x=.4x dollars. The increase of .4x dollars is 50% over the wrong price of .8x dollars. The answer is (E).

Official Problems that Use SAT® Secret #4: Substitution

page 426, problem 12

Let $y=-1$ and $2x=-2,\ x=-1$

(A) $-2x=-2(-1)=2$

(B) $-(2x+y)=-(-2-1)=3$

(C) $2x=2(-1)=-2$

(D) 0

(E) $-y=-(-1)=1$

page 491, problem 13

Let the original price be $100. The price was first increased to $110. And then the new price was decreased by 25% to $82.50. That is 82.5% of the original price, so the answer is (C).

page 522, problem 16

Let $y=3$ and $x=1$.

Then $y^2-x^2=9-1=8$.

(A) $2x=2(1)=2$

(B) $4x=4(1)=4$

(C) $2x+2=2(1)+2=4$

(D) $2x+4=2(1)+4=6$

(E) $4x+4=4(1)+4=8$

page 585, problem 13

Let $n=0.5$.

Then $n^2=0.25$, and $\sqrt{n}=0.71$

Then $n^2<n<\sqrt{n}$, which is answer choice (E).

page 657, problem 17

Let $p=3, r=5, s=7$

Then $n=3\cdot5\cdot7=105$.

The factors of 105 are:

1, 3, 5, 7, 15, 21, 35 and 105.

There are eight of them.

page 672, problem 14

Because x+9 must be a perfect square, try x=7. Check that

$\sqrt{7+9}=4=7-3$. Then substitute 7 into the answers.

(A) $x=x^2, 7\neq49$

(B) $x=x^2+18, 7\neq49+18$

(C) $x=x^2-6x, 7=49-42$

(D)
$x=x^2-6x+9, 7\neq49-42+9$

(E)
$x=x^2-6x+18, 7\neq49-42+18$

SAT® Secret #5: Roman Numerals

Every SAT® can be expected to have one or two Roman numeral problems. They are a variation of the multiple choice problem.

The way to approach a Roman numeral problem is to treat each Roman numeral like a true-false question. Mark each Roman numeral as true or false, then match the pattern of your answers to the answer choice.

EXAMPLE PROBLEM:

Two sides of a triangle are each 5 units long. Which of the following statements MUST BE true?

I. At least two of the angles of the triangle are congruent.

II. The third side of the triangle must be less than 10 units long.

III. The triangle is equilateral.

(A) I only

(B) II only

(C) III only

(D) I and II only

(E) I and III only

SOLUTION:

I. True. If two sides are congruent then the triangle is isosceles and its base angles are congruent.

II. True. The triangle inequality states that any side of a triangle must be less than the sum of the other two sides.

III. False. The triangle could be equilateral, but it might not be.

The answer is (D).

Official Problems that Use SAT® Secret #5: Roman Numerals

page 461, problem 8

I. False. a+1 is even, and an even times an odd is always even.

II. True. a+1 is even, and an even plus an odd is always odd.

III. True. a+1 is even, and an even minus an odd is always odd.

The answer is (E)

page 491, problem 16

Use substitution for this.

Let $x = 2$, then $y = 2.5$

I. True. $2.5 \neq 1$

II. False. 2.5 is not an integer

III. True. $2(2.5) > 2^2, 5 > 4$

The answer is (D).

page 550, problem 14.

Use substitution, let x=0.5

I. True. $(0.5)^2 > (0.5)^3$

II. True. $0.5 > \frac{0.5}{2}$

III. True. $0.5 > (0.5)^3$

The answer is (E).

page 671, proglem 8

I. True. 13 is odd and twice 13 is equal to 26.

II. True. 26 is even and 26 itself is 26.

III. False. 52 is even and 52 itself is 52, not 26.

The answer is (C)

page 795, problem 15

Use substitution.

Let $x = 3$ and $y = 5$.

I. True. $xy = 3 \cdot 5 = 15$ and 15 is divisible by 15.

II. False. $3x + 5y = 3 \cdot 3 + 5 \cdot 5 = 34$ and 34 is not divisible by 15.

III. True. $5x + 3y = 5 \cdot 3 + 3 \cdot 5 = 30$ and 30 is divisible by 15.

The answer is (D).

SAT® Secret #6: Make a List

Sometimes a problem cannot be solved with algebra or substitution, and it is structured so that backsolving won't help. In these cases, it is necessary to make a list of possible answers to find a solution. It is best to structure the list in the form of a table.

Typically, what you have to do is organize the list and then look for a pattern.

EXAMPLE PROBLEM:

If the sum of two distinct prime numbers is odd, the value of one of them must be:

(A) 0

(B) 1

(C) 2

(D) 3

(E) 4

SOLUTION:

Construct a table of prime number and their sums, then look for a pattern.

	2	3	5	7
2	----	5	7	9
3	5	----	8	10
5	7	8	----	12
7	9	10	12	----

Notice the cells of the table which are odd. The only odds appear in the row or column involving 2. The other cells are all even. Therefore the answer is (C).

GUESS: This is a great problem for guessing because 0, 1 and 4 are not prime numbers. If you knew this, you could have guessed either (C) or (D).

Official Problems that Use

SAT® Secret #6: Make a List

page 521, problem 14

We make a table of ordered pairs (x, y):

x	y	2x+3y
1	1	5
1	2	8
2	1	7
2	2	10

We can see that only the first ordered pair satisfies the inequality, so the answer is (A).

page 807, problem 14

We are told that n is an integer greater than 1, so we make a table based on values of n:

n	n+3	n+10
2	5	12
3	6	13
4	7	14

For n=2, we see that there are no common factors of 5 and 12. For n=3, we see that there are no common factors of 6 and 13. For n=4 we see that that there is a common factor of 7. The answer is (B).

page 839, problem 6

We make a table of consecutive positive integers.

n	n/3	remainder
1	1/3	1
2	2/3	2
3	1	0
4	$1\frac{1}{3}$	1
5	$1\frac{2}{3}$	2
6	2	0

The only answer choice that matches the table is (D).

page 859, problem 14

We make a table of prime numbers and their products.

	2	3	5	7	11	13
2	---	---	---	---	---	---
3	6	---	---	---	---	---
5	10	15	---	---	---	---
7	14	**21**	35	---	---	---
11	**22**	33	55	77	---	---
13	**26**	39	65	91		---

Only three products are between 20 and 30. The answer is (D).

SAT® Secret #7: Mark up the Diagram

This is a pretty simple technique, but it is surprising how infrequently students use it.

I tell students over and over again to be aggressive when they see a geometry problem with a diagram. Sometimes I think they consider SAT® diagrams to be sacred or something, and are reluctant to mark them up.

The technique is simple and effective. As you read the problem, mark the diagram to indicate what is given. Then add information to the diagram that can be easily deduced from what is given. At that point, it should be much easier to solve the problem.

EXAMPLE PROBLEM:

In the diagram below, lines *l*, *m*, and *n* are parallel to each other. Find the measure (in degrees) of the angle indicated by *x*.

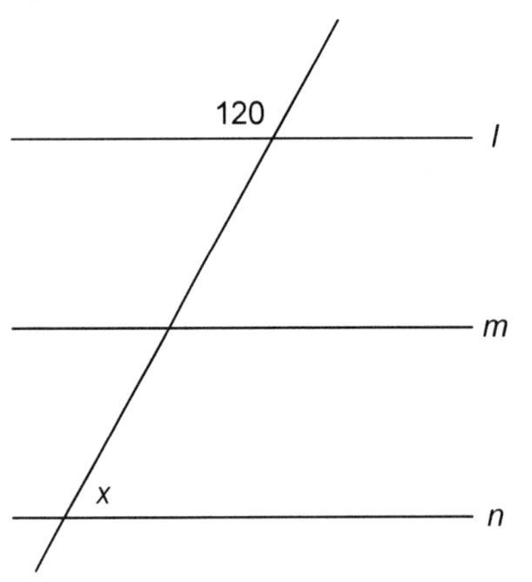

SOLUTION:

This is a relatively easy problem, but it can be made even easier if you mark up the diagram.

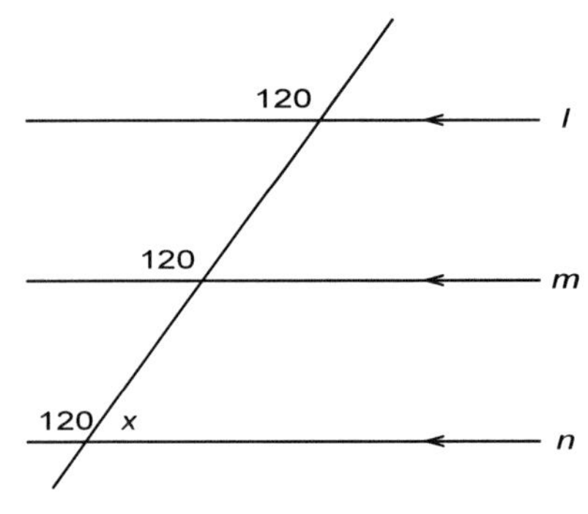

Notice that we have three corresponding angles of 120 degrees. Because angle *x* is supplemental to 120, it must be 60 degrees.

Official Problems that Use

SAT® Secret #7: Mark up the Diagram

page 425, problem 9

We are given that AD is 1 and DC is $\sqrt{3}$. Notice that ΔABD is isosceles, so BD is 1.

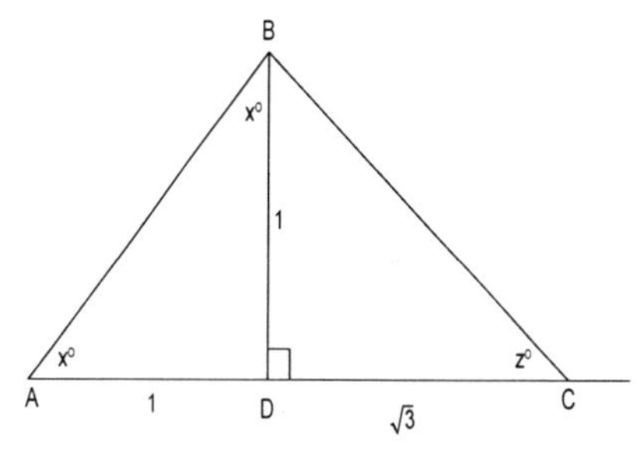

Notice that ΔBCD is 30-60-90 (check the special right triangles in the instructions). So z must be 30. **(D)**

page 462, problem 14

Draw a line along the bottom to see that the triangle is equilateral. Add the five sides together to get 30 units. **(D)**

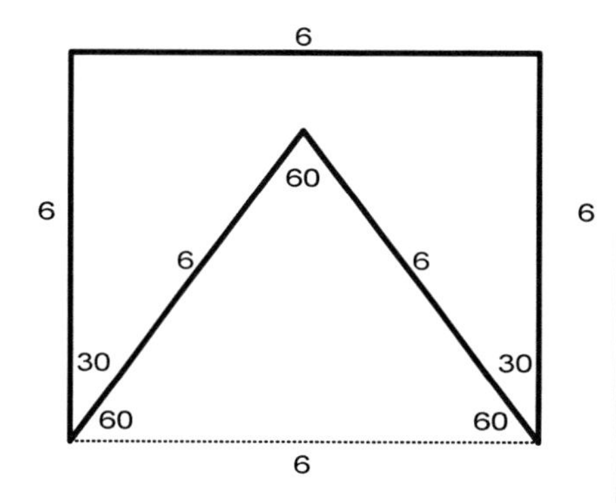

page 520, problem 9

We are given parallel lines with angles as indicated. Notice that the 80 degree angle and the angle consisting of (a+50) are congruent because they are alternate interior angles. So a must be 30. Notice that a and x are supplementary. So x is 180-30=150. **(A)**

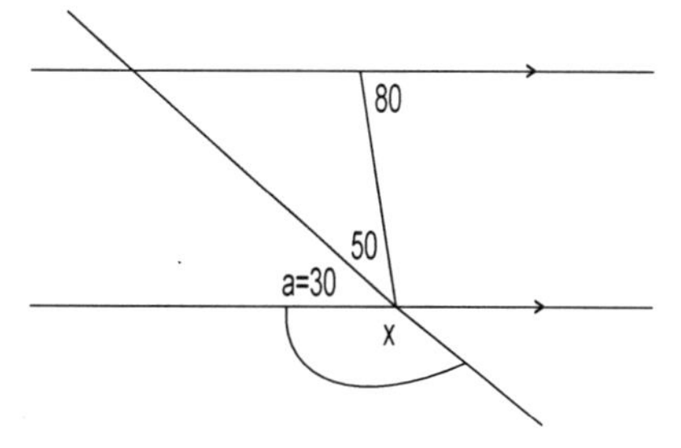

page 744 problem 3

We are given parallel lines with angles as indicated. Notice that a must be 80 because it is corresponding. Likewise b must be 70 because it is corresponding. Therefore z is 180-(80+70)=30. **(A)**

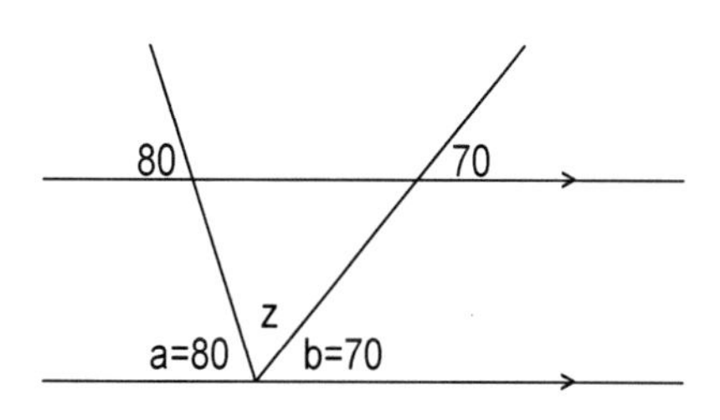

SAT® Secret #8: Averages

For some reason, the folks that write problems for the SAT® like to play with the idea of an average (also known as the arithmetic mean). The secret to remember is that when you see the word "average" on the test, it is best to work with sums.

To see what this means, consider the standard definition of an average:

$$\text{average} = \frac{\text{sum}}{\text{count}}$$

Some problems on the test will use the standard definition. But other problems will use the more difficult definition:

$$\text{sum} = (\text{average}) \cdot (\text{count})$$

An example of the latter is given below, and official examples on the next page use both definitions for the average.

EXAMPLE PROBLEM:

Twelve students took a math test and the average (arithmetic mean) of their scores was 85 points. However, Billy was out sick that day. After Billy took the test, the average score dropped to 83 points. What was Billy's score on the math test?

(A) 24

(B) 59

(C) 83

(D) 85

(E) Cannot be determined

SOLUTION:

Without Billy, the sum of the math scores was 85x12=1020 points.

With Billy, the sum of the math scores was 83x13=1079 points.

Therefore Billy's score on the test was 1079-1020=59, answer choice (B).

Remember to work with the sums!

Official Problems that Use SAT® Secret #8: Averages

page 397, problem 5

Work with sums!

x+y=10 and

x+y+z=3(8)=24.

So 10+z=24,

and z=24-10=14.

(B)

page 522, problem 18

Work with sums! $x+y=2k$.

So $\frac{x+y+z}{3}=\frac{2k+z}{3}$.

(A)

page 614, problem 11

This is a little tricky, but it is just based on the first definition. You do have to remember how to divide fractions, though.

$$k=\frac{sum}{avg}=\frac{sum}{\frac{sum}{count}}$$

$$k=sum\left(\frac{count}{sum}\right)=count$$

(D)

page 670, problem 6.

$$\frac{6+6+12+16+x}{5}=x$$

$$40+x=5x$$

$$40=4x,\ 10=x$$

(D)

page 721, problem 18

First sum = 70p

Second sum = 92n

Set up the average:

$$\frac{70p+92n}{p+n}=86$$

$$70p+92n=86p+86n$$

$$6n=16p$$

$$\frac{6}{16}=\frac{p}{n}$$

page 857, problem 7

$$\frac{x+3x}{2}=12$$

$$\frac{4x}{2}=12$$

$$4x=24,\ x=6$$

(C)

SAT® Secret #9:

Use Your Graphing Calculator

This may be a surprise to you. Certainly, it is not a secret that a graphing calculator is allowed on the SAT®. However, most problems on the test are constructed so that a calculator is not needed at all. When a calculator is needed, it is often to do simple arithmetic.

As a result, what happens to some students is they forget to use their graphing calculator. It can be very useful in problems, especially when substitution is used in combination with the graphing calculator.

EXAMPLE PROBLEM:

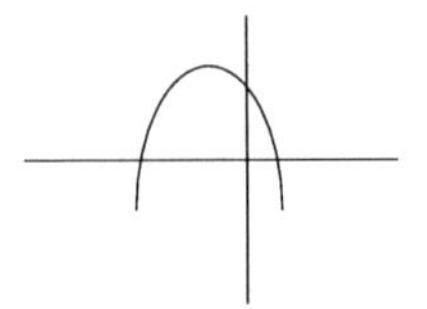

Above is the graph of the function $y = ax^2 + bx + c$.

Which of the graphs below could be the graph of the function $y = -ax^2 - bx - c$?

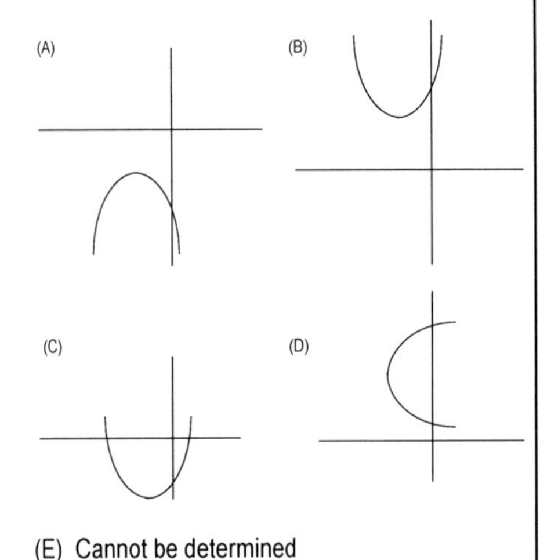

(E) Cannot be determined

SOLUTION:

The A-student might notice that if the original graph is f(x) then the new graph is just -f(x), which is a reflection of the original graph about the x-axis. The new graph is graph (C).

But suppose you did not notice or you do not know the rules for reflections.

You could try values of a, b, and c in your graphing calculator until you have a graph that looks like the one on top. Let's say for example that you end up with $y = -x^2 - 3x + 3$ Enter this and check that it looks like the graph on top. Next put $y = x^2 + 3x - 3$ into your calculator and notice that it looks most like graph (C).

Official Problems that Use SAT®

Secret #9: Use Your Graphing Calculator

page 427, problem 15

Because PQ=6 we know the graphs intersect at x=3.

Put $y = x^2$ into your calculator and then put $y = 12 - x^2$ into the next line. Find that their intersection is at x=2.45. So eliminate choice (C).

Try the other answer choices. You will find that $y = 18 - x^2$ intersects with $y = x^2$ at x=3.

(E)

page 462, problem 12

Put $y = 5x - 10$ into your graphing calculator and see that the line crosses the x-axis at x=2. (D).

page 487, problem 2

Choose particular values of y=mx+b where m is negative and b is positive. Say $y = -2x + 3$. Put this into your graphing calculator. Notice that it most resembles graph (D).

page 490, problem 11

Plug specific values of a, b and c into $y = ax^2 + bx + c$ where a and c are negative and b is positive. Try $y = -x^2 + x - 1$. Notice that its graph most closely resembles graph (A).

page 682, problem 11

Put the equation $y = -\frac{4}{5}x$ into your calculator. Then put in the equation $y = \frac{5}{4}x$. It is not a reflection (slope is too big), so eliminate (A). Next try $y = \frac{4}{5}x$ and see that it is a reflection. (B)

page 683, problem 14

Plug specific values of b and c into $y = x^2 + bx + c$ where b and c are positive. Try $y = x^2 + x + 1$. Notice that its graph most closely resembles graph (E)

.

SAT® Secret #10: Mixture Problems

A mixture problem is a special type of problem that is relatively popular on the test. When you see one for the first time, it may seem difficult to do. But once you get used to it, it is very easy.

Mixture problems can be hard to spot. They may appear on the test as word problems, ratio problems, or geometry problems. A good variety of mixture problems appears on the next page.

EXAMPLE PROBLEM:

After school John works in a pizza parlor, where the boss has a secret recipe for dough. The recipe is 7 parts flour to 2 parts water. If the boss asks John to prepare 54 pounds of dough, how many pounds of flour should he use?

(A) 12
(B) 14
(C) 42
(D) 49
(E) 52

SOLUTION:

To preserve the secret recipe, we know that the ratio of flour to water will always be 7/2, but we do not know the multiplier, shown below as x.

$$\frac{flour}{water} = \frac{7}{2} = \frac{7x}{2x}$$

We also know that the sum of the flour used and the water used will be 54 pounds.

7x+2x=54, x=6.

Therefore John will use 7(6)=42 pounds of flour and 2(6)=12 pounds of water.

(C)

Official Problems that Use SAT® Secret #10: Mixture Problems

page 472, problem 6

This problem is unusual because they give the ratio of eggs and ask about the total. But just do as shown in the previous example and get

2x+3x=5x=total

It is clear that the total number of eggs must be a multiple of 5, and the only answer choice that is not a multiple of 5 is (B).

page 488, problem 6

This is a geometry form of mixture problem. Just set up the equation

2x+3x+4x=360

9x=360, x=40. (C)

page 804, problem 4

From probability:

$$\frac{2}{5}=\frac{\text{apples}}{\text{apples + non-apples}}$$

So the ratio of apples to non-apples must be 2:3

2x+3x=5x=total

It is clear that the total number of fruit must be a multiple of 5, and the only answer choice that is not a multiple of 5 is (C).

page 843, problem 16

The ratio of peanuts to cashews is:

$$\frac{p}{c}=\frac{5}{2}=\frac{5x}{2x}$$

$$5x+2x=4$$

$$x=\frac{4}{7}$$

$$c=2x=2\left(\frac{4}{7}\right)=\frac{8}{7}$$

page 870, problem 12

This is another mixture problem in the form of geometry. We know that the angles are 2:3:4 and they must sum to 180 degrees.

$$2x+3x+4x=180$$

$$9x=180,\ x=20$$

$$2x=40,\ 4x=80$$

$$80-40=40$$

(C)

SAT® Secret #11: Draw a Picture

This one is like the calculator secret – it is not a secret at all, but so few students do it! Sometimes they forget, but mostly they think that it will use up too much time. Although it may feel like it uses a lot of time, it really does not, and it will reduce the likelihood of making a very silly mistake that you will regret.

It is especially helpful to draw a picture with word problems.

EXAMPLE PROBLEM:

Sally drives south at the rate of 30 mph and at the same time Willy drives west at the rate of 60 mph. If Sally and Willy started at the same place at the same time, how far apart will they be after two hours?

(A) 104 miles

(B) 134 miles

(C) 180 miles

(D) 10,800 miles

(E) 18,000 miles

SOLUTION:

After two hours Sally will have driven 60 miles and Willy will have driven 120 miles, so the picture looks like this:

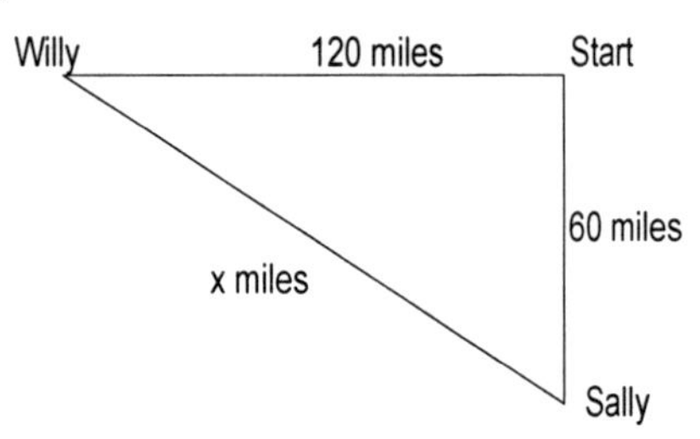

It is now easy to see that we are interested in the hypotenuse of the right triangle.

$60^2 + 120^2 = x^2$

$18,000 = x^2$

$134 = x$

(B)

Official Problems that Use
SAT® Secret #11: Draw a Picture

page 427, problem 16

For this problem, a particular type of picture is a good idea – the Venn diagram.

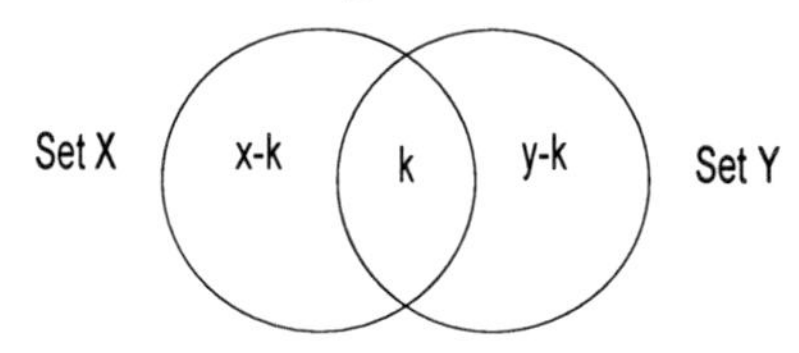

With the diagram in hand, it is easy to see:

z=(x-k)+(y-k)=x+y-2k

(D)

page 549, problem 8

A quick sketch makes it easy to see that the radius is 5.

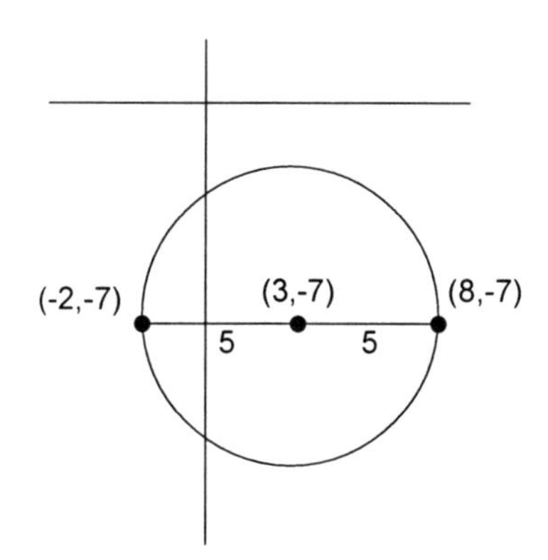

The answer is (E).

page 613, problem 7

A sketch clarifies this one easily and avoids the trap of answer choice (B), x=5.

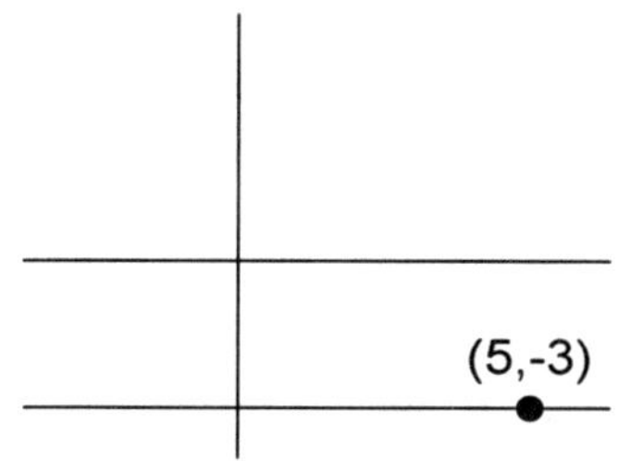

The answer is (C), y=-3.

page 776, problem 8

We can pick any slope we want, so we choose slope=1 and draw the rectangle. Remember that slopes of perpendicular lines are negative reciprocals.

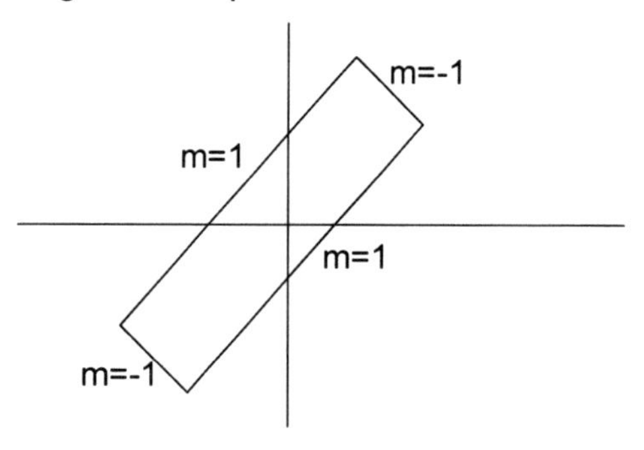

The product is (1)(-1)(1)(-1)=1.

The answer is (D).

SAT® Secret #12: Area Subtraction

In Geometry problems it is common to be asked to find the area of a strange-looking shape. Sometimes it is useful to divide the strange shape into familiar shapes and sum the areas of the familiar shapes. On the SAT®, it is more common to find the area of the strange shape by subtracting familiar shapes.

This may sound confusing until you see a few problems.

EXAMPLE PROBLEM:

Two congruent circles are inscribed in a rectangle, as shown below. Find the area (in square feet) of the shaded region if the radius of each circle is 5 feet.

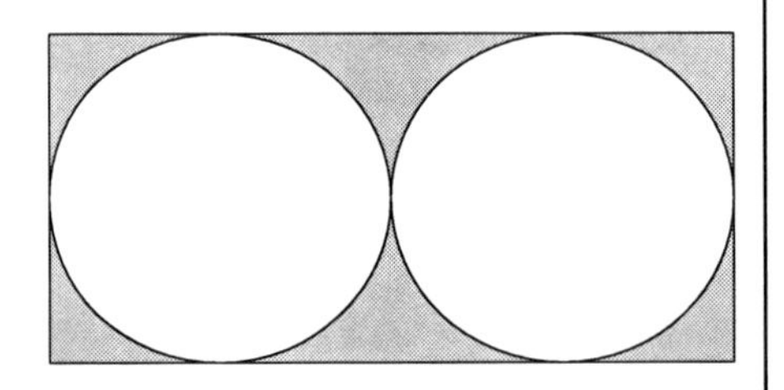

(A) $50-25\pi$

(B) $50-50\pi$

(C) $200-10\pi$

(D) $200-25\pi$

(E) $200-50\pi$

SOLUTION:

If each radius is 5 then each diameter is 10. So the rectangle is 10x20=200 sq ft.

Each circle is $\pi 5^2 = 25\pi$ sq ft. So the shaded region is $200-50\pi$ square feet.

The answer is (E).

Notice the answers that correspond to silly mistakes. (A) and (B) assume that you calculated the area of the rectangle incorrectly. (C) assumes you forgot to square the 5. (D) assumes you forgot to allow for two circles instead of one.

Official Problems that Use
SAT® Secret #12: Area Subtraction

page 397, problem 6

The area of the big square is 25 and the area of the little white square is x^2. So the area of the shaded region is $25 - x^2$. (E)

page 585, problem 16

If the radius of each white quarter circle is 1 unit, then the rectangle must be 2x1=2 square units. The area of a full circle would be $\pi 1^2 = \pi$, but we have a half circle (two quarter circles), so their combined area is $\frac{1}{2}\pi$. The area of the shaded region is $2 - \frac{1}{2}\pi$. (B)

page 657, problem 16

This one is a little trickier. We want the area of the large circle. The area of any circle is πr^2. The shaded area is the area of the large circle less the area of the white circle. So we can set up the equation:

$$64\pi = \text{large circle} - \pi 6^2$$

$$64\pi + 36\pi = 100\pi = \text{large circle}$$

If the area of the large circle is 100π then its radius must be 10.

page 721 problem 16

The area of the big square is 3x3=9 square units.

Notice the four white right triangles in the corners. Each has a leg of 1 unit and another leg of 2 units. So each right triangle has an area of $\frac{1}{2}(2)(1) = 1$ square unit. The area of the shaded region is 9-4=5 square units.

SAT® Secret #13: Geometry Guessing

On some problems you can take advantage of the fact that all diagrams on the test are drawn to scale unless indicated otherwise. Sometimes you can use your eyes to eliminate two or more of the answer choices and take a guess.

Even when a diagram is not drawn to scale, you can try to draw your own diagram to scale and take a guess based on that.

EXAMPLE PROBLEM:

In the diagram below, what is the measure of angle X (in degrees)?

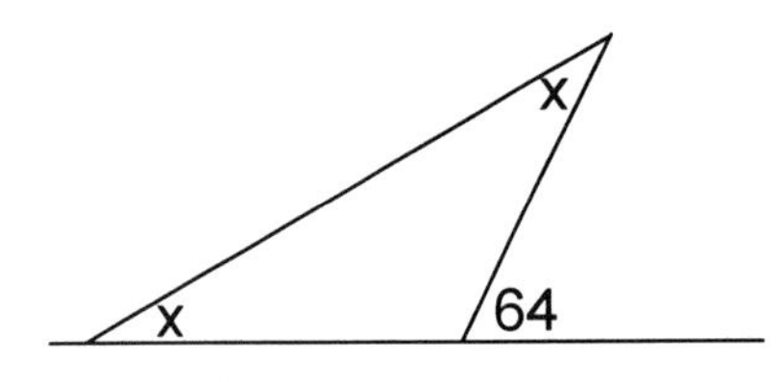

(A) 12

(B) 22

(C) 32

(D) 42

(E) 64

SOLUTION:

The angle supplemental to the 64 degree angle is 116 degrees. The angles of a triangle must sum to 180, so we have the equation

$x+x+116=180$

$x=32$

The answer is (C).

Suppose you did not see how to solve the problem. It is fairly clear that the angle X is smaller than 64 degrees. (It is a little less clear that the angle X is less than 42 degrees, but a fairly safe bet.) It is also fairly clear that angle X is more than 12 degrees. At this point, having eliminated answer choices (A) and (E), or (D) and (E), you should take a guess from the remaining three answer choices.

Official Problems that Use SAT® Secret #13: Geometry Guessing

page 462, problem 14

Suppose you can't solve for the lengths of those diagonal lines. Without them, the perimeter is 18. The lines look congruent to eachother, so they would each have to be 3 units for the perimeter to be 24. Clearly the lines are longer than 3 units and so 24 is too low. Just as clearly, 25 is too low also, thereby eliminating (A) and (B). Take a guess (answer is D).

page 612, problem 6

Angle r looks more like 70 degrees, so re-draw this one.

We know that s+t+u plus that blank angle sum to 360 degrees because they form a circle. The blank angle looks to be more than 90 degrees, say at least 100 degrees. Then

$s+t+u+100^{+}=360$

$s+t+u=260$ or less.

We can safely eliminate (E) and maybe (D), then take a guess. The answer is (A), making the blank angle 130 degrees, which looks reasonable.

page 671, problem 7

Take a long look at angle Z. It clearly is larger than the 40 degree angle shown, and does not look like it is near 75 degrees. Take a guess after eliminating (A) and (E). The answer is (C).

page 672, problem 12

Answer choices (C) and (D) have positive slopes. The only answer choices with a negative slope are (A), (B), and (E). So take a guess. The answer is (E).

page 806, problem 10

Take a long look at angle X. It is clearly less than 84 degrees. You could guess that it is also larger than 60 degrees, or less than 80 degrees, whichever suits your eye. Then guess from the remaining choices. The answer is (C).

Tracking Log for Practice SAT® Tests

Date	Test	Correct	Wrong	Skipped	Raw Score*	Scaled Score**

*The sum of correct, wrong and skipped problems should be 54. The raw score is one point for every correct problem minus one-quarter point for every wrong multiple choice problem.

**Find the scaled score from the grid at the end of each practice test in The Official SAT Study Guide. Take the midpoint of the range that is given for your raw score.

ACT® Secret #1: Read-Solve-Guess

This technique applies to every problem and every type of student, so pay close attention.

Read

Read the problem carefully, moving as quickly as you can without sacrificing comprehension. If you know how to solve the problem, go ahead and solve it (see below). If you do not know how to solve the problem, then guess (see below).

As you are reading, you may decide to "skip" (randomly guess) the problem because it is too long or it involves math you do not know. Also, it is likely that you will not have enough time to read all of the 60 problems on the test in the 60 minutes allowed. Depending on your target score, it is ok not to read some number of problems (for details, see Secret #2: Find Your Pace). For problems that are not read, bubble-in an answer choice at random and move on.

Solve

The best way to answer a problem is through a "direct solution," where you read the problem carefully and calculate your answer using the math learned in school. Match your answer to one of the answer choices and bubble that in.

Guess

If you have no understanding of the problem, randomly bubble-in an answer choice.

Otherwise try to eliminate some answer choices before taking a guess. Often, you should be able to eliminate at least one choice. For more details, see Secret #3: Educated Guessing.

A wrong answer counts the same as a blank, so always bubble-in an answer choice for every question. Mark your question book so that you can come back to the problem if you have time remaining at the end.

More Information on ACT® Secret #1: Read-Solve-Guess

Below is a summary of the read-solve-guess technique:

- For each problem, either read or skip.
- If read, try to solve the problem. If you are unable to solve it, try to eliminate some answer choices and then bubble-in an educated guess. If you cannot eliminate any answer choices, randomly bubble-in an answer.
- If skip, remember to bubble-in any answer choice.

ACT® Secret #2: Find Your Pace

The objective of this technique is to figure out how much you can shorten the test and still reach your goal. By shortening the test, I mean pacing yourself so that you do not have enough time to read some number of problems at the end. Keep in mind that most students cannot read-solve-guess 60 math problems in 60 minutes without making lots of mistakes. Shortening the test lets you set a pace that will achieve your goal without driving you crazy.

Although you want to move as quickly as possible, if you go too fast you will make too many mistakes to reach your goal. If you go too slowly, you will not attempt enough problems to reach your goal.

For some students, It is very hard to become comfortable with shortening the test. Years of experience in school have taught you that you should try every problem on the test because you may at least get partial credit. Also, if you skip the more difficult problems toward the end of the test you cannot possibly expect a good grade because that's where most of the points are. You must unlearn this!

Every problem on the ACT® is worth one point, no matter how hard it is. Although easy problems appear throughout, the later parts of the test tend to have fewer easy problems. Also, early math (e.g., fractions and percentages) tends to come early in the test and later math (e.g., trigonometry and logarithms) tends to come later.

Shortening the test is the best way to cope with the ACT®. You can reach your goal by spending enough time on enough problems, without attempting every problem. Some **approximate** guidelines are given below. Try them and see how you do. Make adjustments to your pace after every practice test.

If your target score is 20, you could read-solve-guess the first 32 problems and "skip" (randomly guess) the remaining 28 problems (47% of the test!). If your target score is 25, you could read-solve-guess the first 40 problems and "skip" (randomly guess) the remaining 20 problems (33% of the test!). If your target score is 30+, you could read-solve-guess to the first 54+ problems and "skip" (randomly guess) the remaining 6 or fewer problems.

ACT® Secret #3: Educated Guessing

We now begin techniques with specific examples. The technique of educated guessing applies to problems that you have read and understood to some degree, but cannot solve. In that situation, try to eliminate at least one answer choice before guessing.

If you have not read the problem or if you have no idea what the problem is about, randomly bubble-in one of the answer choices and move on.

As stated before and will be stated again, the best approach is the direct solution. Read the problem carefully, calculate the solution, and match your answer with an answer choice. If you are not able to calculate the solution, then try educated guessing or make a random guess. Don't forget to always bubble-in some answer.

EXAMPLE PROBLEM:

If a math class consists of 21 students total and there are five more boys than girls, how many girls are in the class?

(A) 5
(B) 8
(C) 11
(D) 13
(E) 16

AN EDUCATED GUESS:

Even if you cannot solve this problem directly, (backsolving is a good approach, see Secret #4), you certainly can eliminate some answer choices before taking a guess.

We are told that there are more boys than girls in the class, so girls cannot comprise more than half the class. This simple observation enables you to eliminate answer choices (C), (D) and (E). Take a guess from the remaining choices. The answer is (B), 8 girls and 13 boys.

Real Problems that Use

ACT® Secret #3: Educated Guessing

page 164, problem 2

Even if you forgot your laws of exponents, you can still multiply. Just multiply $3 \cdot 2 \cdot 4 = 24$. Eliminate answer choices (F) and (G), then guess. The answer is (H).

page 166, problem 12

Even if you forgot how to do probability, it is clear that less than half the marbles in the bag are white, so the probability of choosing a marble that is not white must be greater than one-half. Eliminate (J) and guess. The answer is (K).

page 174, problem 41

Add the sides that are marked. Their sum is 4+6+4+10+6+26=56. So the perimeter must be greater than 56. Eliminate answer choices (A), (B), and (C), then guess. The answer is (E).

page 176, problem 52

Drawing and then counting all of the diagonals could be messy and drive you crazy. But there are clearly more than eight diagonals total. There are five diagonals just from one vertex. Eliminate (F) and guess. The answer is (H).

page 306, problem 11

If one hamburger and one soft drink cost $2.10, it is safe to assume that the soft drink cost less than $1.00 (sugar water costs less than ground meat). Eliminate (D) and (E), then guess. The answer is (C).

page 453, problem 15

Even if you forgot how to calculate the midpoint, it must be between -5 and 17. Eliminate (A) and (E), then guess. The answer is (B).

page 457, problem 31

Use your calculator to find that $(-2)^4 = 16$. Eliminate (A), (B), and (C), then take a guess. The answer is (E).

ACT® Secret #4: Backsolving

Many students prefer multiple choice problems because the answer is right there. It is among the five answer choices and "all you have to do" is find it. One way to find it is to backsolve.

As stated before, the best approach is the direct solution. Read the problem carefully, calculate the solution, and match your answer with an answer choice. If you are not able to calculate the solution, then backsolving is an option.

To backsolve a problem, simply plug each of the answer choices into the original problem and see which one works. Many times the answer choices are ordered from smallest to largest (or largest to smallest). If that is the case, try the middle answer choice first. If it is not correct, go up or down from there.

Some problems are structured so that they cannot be backsolved, so this technique cannot be applied throughout the test. Also, backsolving takes time.

EXAMPLE PROBLEM:

What value of x satisfies the equation $2|x-3|=8$?

(A) -7
(B) -1
(C) 0
(D) 1
(E) 2

SOLVE BY BACKSOLVING:

Plug in answer choices.

(A) $2|-7-3|=2\cdot 10=20$

(B) $2|-1-3|=2\cdot 4=8$

(C) $2|0-3|=2\cdot 3=6$

(D) $2|1-3|=2\cdot 2=4$

(E) $2|2-3|=2\cdot 1=2$

The answer is (B).

Real Problems that Use
ACT® Secret #4: Backsolving

page 165, problem 8

(F)
$$4\left(\frac{7}{5}\right)+3=\frac{28}{5}+3=\frac{43}{5}$$
$$9\left(\frac{7}{5}\right)-4=\frac{63}{5}-4=\frac{43}{5}$$

(G)
$$4\left(\frac{5}{7}\right)+3=\frac{20}{7}+3=\frac{41}{7}$$
$$9\left(\frac{5}{7}\right)-4=\frac{45}{7}-4=\frac{17}{7}$$

(H)
$$4\left(\frac{7}{13}\right)+3=\frac{28}{13}+3=\frac{67}{13}$$
$$9\left(\frac{7}{13}\right)-4=\frac{63}{13}-4=\frac{11}{13}$$

(J)
$$4\left(\frac{1}{5}\right)+3=\frac{4}{5}+3=\frac{19}{5}$$
$$9\left(\frac{1}{5}\right)-4=\frac{9}{5}-4=-\frac{11}{5}$$

(K)
$$4\left(\frac{-1}{5}\right)+3=\frac{-4}{5}+3=\frac{11}{5}$$
$$9\left(\frac{-1}{5}\right)-4=\frac{-9}{5}-4=-\frac{29}{5}$$

page 453, problem 18

First, find that $8^3=512$.

(F) $(2^2)(4)=16$

(G) $(2^3)(4)=32$

(H) $(2^4)(4)=64$

(J) $(2^{4.5})(4)=90.51$

(K) $(2^7)(4)=512$

page 456, problem 26

(F) $\frac{7-11}{9-11}=\frac{-4}{-2}=2$

(G) $\frac{7-5}{9-5}=\frac{2}{4}=\frac{1}{2}$

(H) $\frac{7-2.5}{9-2.5}=\frac{4.5}{6.5}=0.7$

(J) $\frac{7-1\frac{2}{3}}{9-1\frac{2}{3}}=\frac{\frac{16}{3}}{\frac{22}{3}}=\frac{16}{22}=\frac{8}{11}$

(K) $\frac{7+5}{9+5}=\frac{12}{14}=\frac{6}{7}$

ACT® Secret #5: Markup the Diagram

This is a pretty simple technique, but it is surprising how infrequently students use it.

I tell students over and over again to be aggressive when they see a geometry problem with a diagram. Sometimes I think they consider ACT® diagrams to be sacred or something, and are reluctant to mark them up.

The technique is simple and effective. As you read the problem, mark the diagram to indicate what is given. Then add information to the diagram that can be easily deduced from what is given. At that point, it should be much easier to solve the problem. Remember that diagrams are not necessarily drawn to scale, though many are.

EXAMPLE PROBLEM:

In the diagram below, lines *l*, *m*, and *n* are parallel to each other. Find the measure (in degrees) of the angle indicated by *x*.

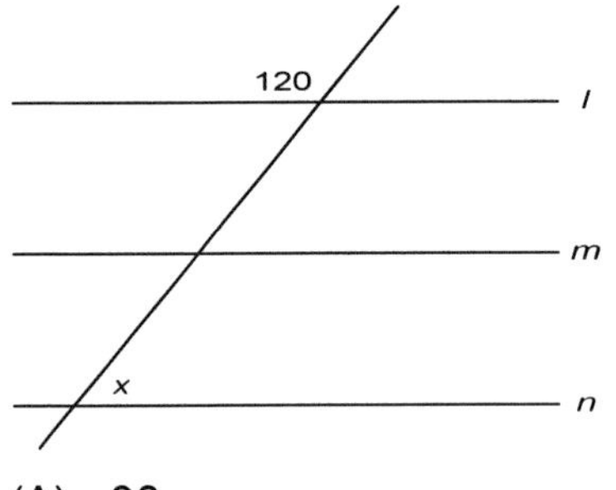

(A) 30

(B) 45

(C) 60

(D) 90

(E) 120

SOLUTION:

This is a relatively easy problem, but it can be made even easier if you mark up the diagram.

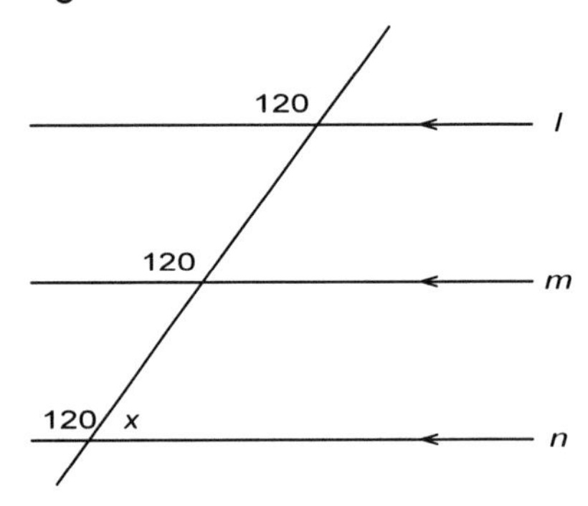

Notice that we have three corresponding angles of 120 degrees. Because angle *x* is supplemental to 120, it must be 60 degrees **(C)**.

Real Problems that Use

ACT® Secret #5: Markup the Diagram

page 169, problem 25

From the diagram it is clear that the vertical radius is the perpendicular bisector of the chord. Once you mark half the chord as 12 units, the solution is simple Pythagorean Theorem $5^2 + 12^2 = 13^2$. **(D)**

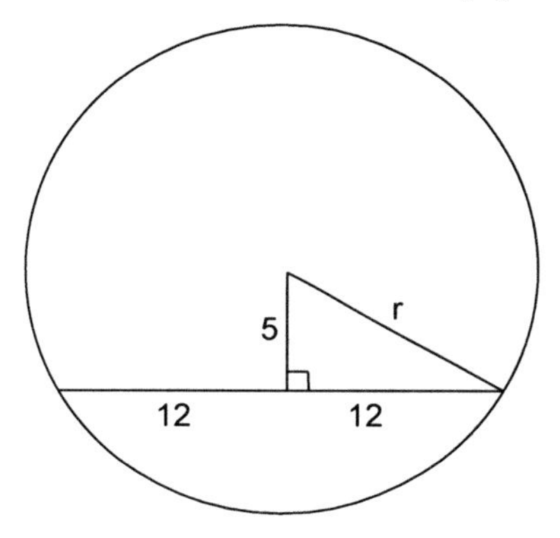

page 171, problem 30

If AD is 30 and AC is 16, then CD must be 30-16=14 units. Mark it. If BD is 20 and CD is 14, then BC must be 20-14=6 units. **(B)**

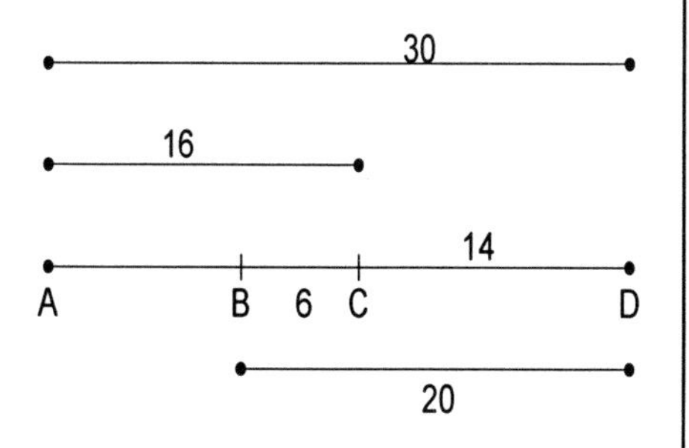

page 173, problem 40

Trapezoids have parallel bases. $\angle CBD \cong \angle BDA$ because they are alternate interior angles. Once $\angle BDA$ is marked as 30, the problem becomes easy, $30 + x + 105 = 180,\ x = 45$. **(K)**

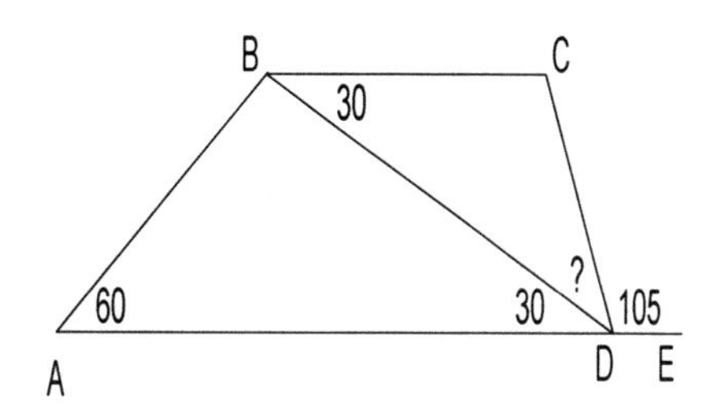

page 458, problem 35

Notice that $140 + \angle DCE = 180$. Mark $\angle DCE$ as 40. Second, notice that $40 + 80 + \angle DEC = 180$. Mark $\angle DEC$ as 60 and also its vertical angle, $\angle FEG$ is 60. Third, mark $\angle FGE$ as 80. The rest is easy. **(A)**

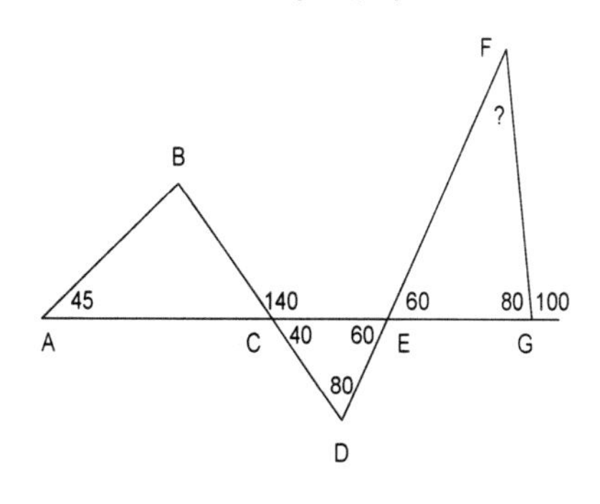

ACT® Secret #6: Draw a Picture

This one is not a secret at all, but so few students do it! Sometimes they forget, but mostly they think that it will use up too much time. Although it may feel like it uses a lot of time, it really does not, and it will reduce the likelihood of making a very silly mistake that you will regret.

It is especially helpful to draw a picture with word problems.

EXAMPLE PROBLEM:

Sally drives south at the rate of 30 mph and at the same time Willy drives west at the rate of 60 mph. If Sally and Willy started at the same place at the same time, how far apart will they be after two hours?

(A) 104 miles
(B) 134 miles
(C) 180 miles
(D) 10,800 miles
(E) 18,000 miles

SOLUTION:

After two hours Sally will have driven 60 miles and Willy will have driven 120 miles, so the picture looks like this:

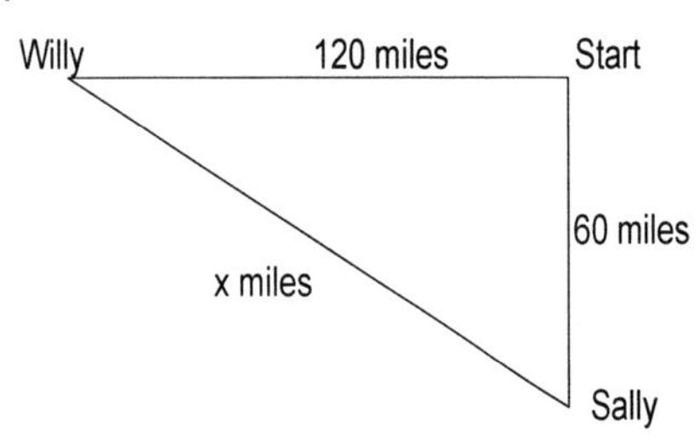

It is now easy to see that we are interested in the hypotenuse of the right triangle.

$60^2 + 120^2 = x^2$

$18,000 = x^2$

$134 = x$

(B)

Real Problems that Use
ACT® Secret #6: Draw a Picture

page 165, problem 6

A picture isn't really necessary but it helps avoid silly mistakes. The perimeter is 150+200+150+200=700. **(J)**

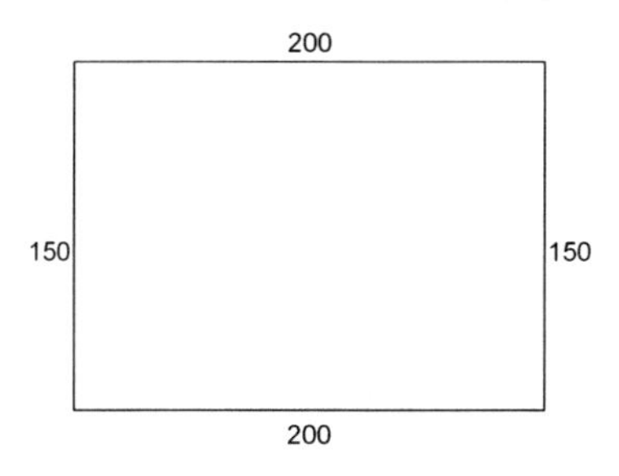

page 173, problem 37

A quick sketch makes it easy to see that this is simple Pythagorean Theorem.

$10^2 + x^2 = 30^2$, $x = 28.3$. **(C)**

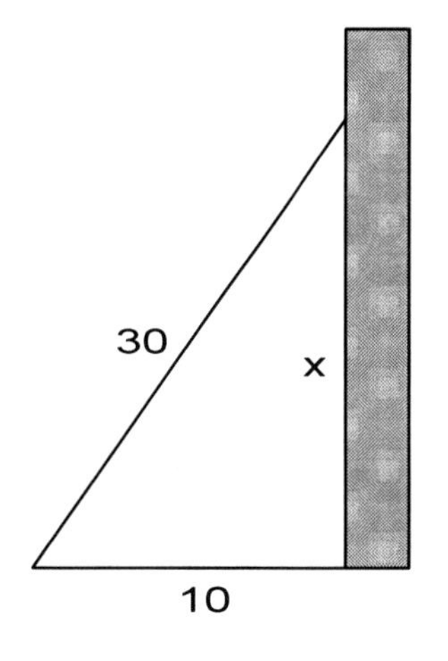

page 307, problem 16

A picture not only clarifies this one, it can actually give you the answer. y=4. **(H)**

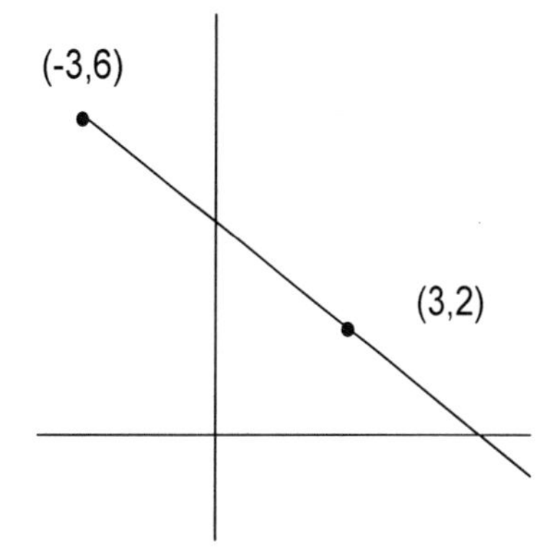

page 457, problem 32

A picture clarifies the sequence of points in this word problem. We make them equidistant because the problem does not specify distances. With the picture in hand, it is clear that the answer is CD<BC. **(K)**

ACT® Secret #7: SOHCAHTOA

For some reason, the folks that write problems for the ACT® like to play with the basic trigonometric functions, sine, cosine and tangent. Less frequently you will see their reciprocals, cosecant, secant, and cotangent, respectively. The basic trig functions are best remembered by using the acronym, SOHCAHTOA.

S = $\sin\theta$ is equal to
O= opposite over
H= hypotenuse.
C= $\cos\theta$ is equal to
A= adjacent over
H= hypotenuse.
T= $\tan\theta$ is equal to
O= opposite over
A= adjacent.

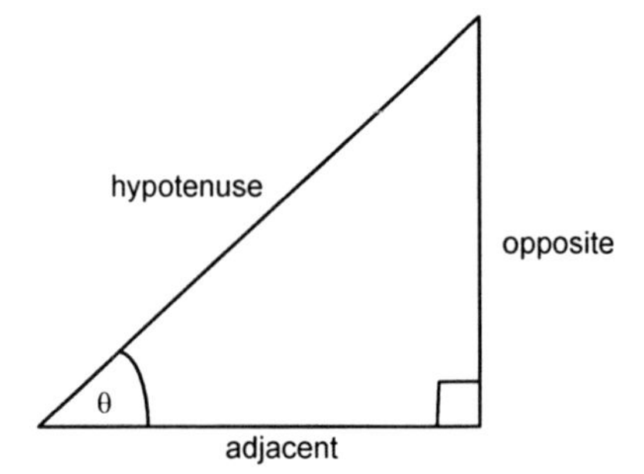

EXAMPLE PROBLEM:

What is the value of x-y in the triangle below?

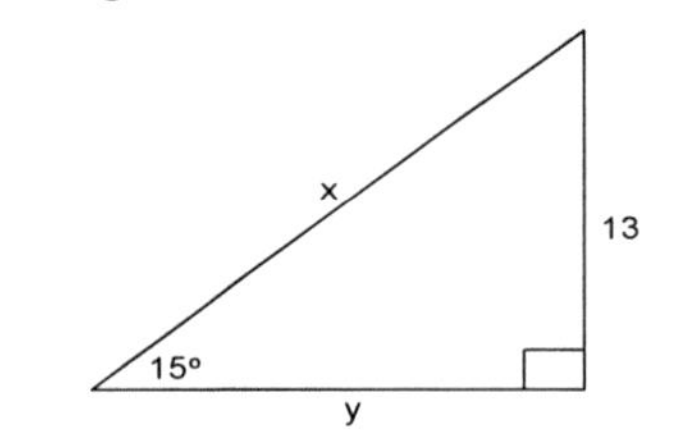

(A) 2
(B) 13
(C) 35
(D) 48
(E) 50

SOLUTION:

Using the graphing calculator in degree mode, we begin by finding x through the equation

$$\sin(15) = \frac{13}{x},\ x = 50$$

The value of y can be found in a similar way:

$$\tan(15) = \frac{13}{y},\ y = 48$$

So x-y = 50-48 = 2.

(A)

Real Problems that Use
ACT® Secret #7: SOHCAHTOA

page 170, problem 28

$$\sin \angle P = \frac{3}{5} = \frac{opposite}{hypotenuse} = \frac{QR}{16}$$

Next cross-multiply and solve:

$$QR = \left(\frac{3}{5}\right)16 = 9.6$$ **(G)**

page 177, problem 54

We are told that θ is in the third quadrant, so the sketch:

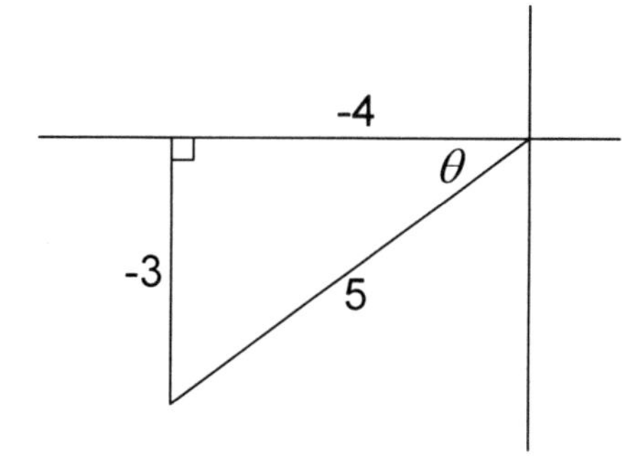

$$\tan\theta = \frac{opposite}{adjacent} = \frac{-3}{-4} = \frac{3}{4}$$ **(J)**

page 314, problem 45

Remember that the cotangent is the reciprocal of the tangent.

$$\cot A = \frac{adjacent}{opposite} = \frac{\sqrt{4-x^2}}{x}$$ **(E)**

page 461, problem 46

$$\tan A = \frac{opposite}{adjacent} = \frac{a}{b}$$

$$\sin B = \frac{opposite}{hypotenuse} = \frac{b}{c}$$

$$(\tan A)(\sin B) = \left(\frac{a}{b}\right)\left(\frac{b}{c}\right) = \frac{a}{c}$$

(F)

page 462, problem 49

We are told that angle A is in the first quadrant, so the sketch:

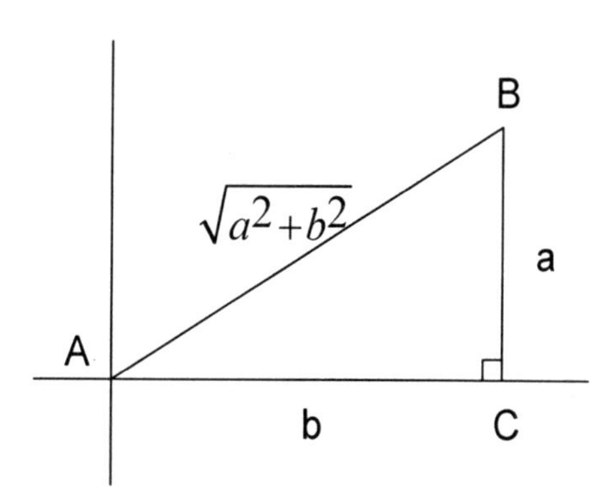

$$\cos A = \frac{adjacent}{hypotenuse} = \frac{b}{\sqrt{a^2+b^2}}$$

(D)

ACT® Secret #8: Use Your Graphing Calculator

This may be a surprise to you. Certainly, it is not a secret that a graphing calculator is allowed on the math section of the ACT® (be aware that the TI-89 is <u>not</u> allowed). However, most problems on the test are constructed so that a calculator is not needed at all. When a calculator is needed, it is often to do simple arithmetic.

As a result, what happens to some students is they forget to use their graphing calculator.

EXAMPLE PROBLEM:

Which of the graphs below could be the graph of the function $y = -x^2 - 3x + 3$?

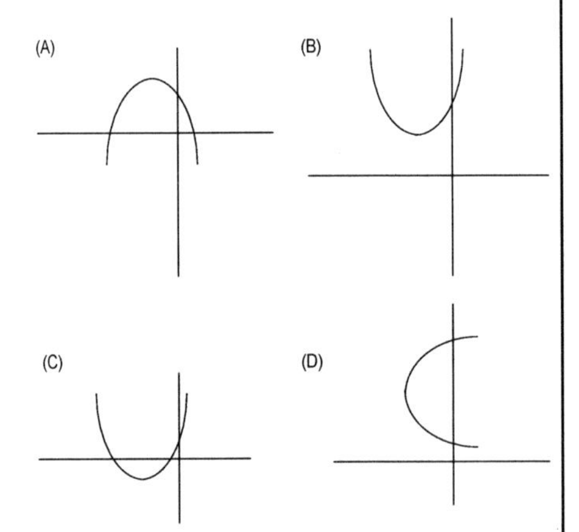

(E) Cannot be determined

SOLUTION:

The A-student would know that the graph is a parabola and that when the leading coefficient is negative, the parabola opens down. That eliminates (B), (C) and (D). That same A-student would know that when x=0, y=3. That matches answer choice (A).

Don't be scared away from problems with parabolas. To answer this question, you do not need to know anything at all about them. Just punch the equation into the graphing calculator and hit the graph key. What comes up on the screen is a graph that matches answer choice (A).

Real Problems that Use ACT® Secret #8: Use Your Graphing Calculator

page 169, problem 23

Enter $x^2 - 36x$ into the calculator and see where the graph crosses the x-axis. You have to adjust the window to see that the graph crosses at x=0 and x=36, or use the calc-zero function. **(B)**

page 178, problem 57

Enter $(2x^2 + x)/x$ into the calculator and hit the graph key. You can immediately see it is a line, so eliminate (D) and (E). The y-intercept is above the origin, so eliminate (B). The x-intercept is greater than -1, so eliminate (C). The answer is **(A)**.

Page 179, problem 60

The author wants you to use the difference formula from trigonometry, but it is not necessary. With your calculator in radian mode, find that $\sin\left(\frac{\pi}{12}\right) = 0.2588$. Use your calculator to convert each of the answer choices to decimal. The answer that matches 0.2588 is **(K)**.

page 308, problem 19

Enter each of the answer choices into the calculator and hit the graph key. All of them are lines except answer **(E)**.

page 317, problem 54

Enter the first equation into the calculator – the other two equations are clearly lines. The graph of $x^2 - 2$ is a parabola opening up, so eliminate (F), and (G). The parabola does not go through the origin, eliminate (H). (J) is not the graph of a function (look at $0<x<1$). The answer is **(K)**.

page 318, problem 55

With the calculator in radian mode, find that $\cos^{-1}(-0.385) = 1.97$. The value of 1.97 is between $\pi/2$ and $2\pi/3$. **(D)**

page 463, problem 50

It will take a while, but you could plug three equations into the calculator for each of the answer choices. **(F)**

ACT® Secret #9: Substitution

Substitution can be widely used, especially instead of using algebra. In fact, unless you are an A-student in math, I strongly suggest solving problems directly using substitution whenever possible. The A-students can go ahead and use algebra. A good example appears below.

EXAMPLE PROBLEM:

A store clerk was asked to markup the price of a pair of shoes by 20%. Instead the clerk marked the price down by 20%. By what percentage does the price now have to be increased in order to be correct?

(A) 20%

(B) 30%

(C) 33%

(D) 40%

(E) 50%

SOLVE BY SUBSTITUTION:

Assume the original price was $100. It was erroneously marked down to $80 when it should have been marked up to $120. Therefore the wrong $80 price must be increased by $40 to obtain the correct price of $120. This is a 50% increase over the wrong price. The answer is (E).

SOLVE BY ALGEBRA:

Assume the original price was x dollars. It was erroneously marked down to .8x dollars when it should have been marked up to 1.2x dollars. Therefore it must be increased by 1.2x -.8x=.4x dollars. The increase of .4x dollars is 50% over the wrong price of .8x dollars. The answer is (E).

Real Problems that Use ACT® Secret #9: Substitution

page 169, problem 21

Let $x=1, y=2, z=3$. Then $\frac{x}{y}=\frac{1}{2}$. Substitute into each of the answer choices.

(A) $\frac{x\cdot z}{y\cdot z}=\frac{1\cdot 3}{2\cdot 3}=\frac{3}{6}=\frac{1}{2}$

(B) $\frac{x\cdot x}{y\cdot y}=\frac{1\cdot 1}{2\cdot 2}=\frac{1}{4}$

(C) $\frac{y\cdot x}{x\cdot y}=\frac{2\cdot 1}{1\cdot 2}=\frac{2}{2}=1$

(D) $\frac{x-z}{y-z}=\frac{1-3}{2-3}=\frac{-2}{-1}=2$

(E) $\frac{x+z}{y+z}=\frac{1+3}{2+3}=\frac{4}{5}$

page 319, problem 60

Let the original 1990 distance be 100 feet. Then the 1991 distance would be 10% more, or 110 feet. Next, the 1992 distance would be 20% more than 110 feet, or 132 feet. The percentage increase from 1990 to 1992:

$\frac{132-100}{100}=32\%$ **(F)**

page 464, problem 56

Choose values of a, b, and c so that:

$a=2b+3c-5$.

Let $b=1, c=2$. Then $a=3$.

The new value of b becomes 0 and the new value of c becomes 4. Now

$a=2(0)+3(4)-5=7$.

The value of a increased from 3 to 7, and increase of 4. **(F)**

page 465, problem 57

Let the edge of the smaller cube be one unit. Then the volume of the smaller cube is $s^3=1^3=1$. The edge of the larger cube must be twice as long, so it must be two units. The volume of the larger cube is $s^3=2^3=8$. The ratio of the volume of the larger cube to the volume of the smaller cube is $\frac{8}{1}=8$. **(D)**

ACT® Secret #10: Averages

The folks that write problems for the ACT® like to work the idea of an average (also known as the arithmetic mean). The secret to remember is that when you see the word "average" on the test, it is often better to work with sums.

To see what this means, consider the standard definition of an average:

$$\text{average} = \frac{\text{sum}}{\text{count}}$$

Some problems on the test will use the standard definition. But other problems will use the more difficult definition:

$$\text{sum} = (\text{average}) \cdot (\text{count})$$

An example of the latter is given below, and real examples on the next page use both definitions for the average.

EXAMPLE PROBLEM:

Twelve students took a math test and their average score was 85 points. However, Billy was sick that day. After Billy took the test, the average score dropped to 83 points. What was Billy's score on the math test?

(A) 24

(B) 59

(C) 83

(D) 85

(E) Cannot be determined

SOLUTION:

Without Billy, the sum of the math scores was 85x12=1020 points.

With Billy, the sum of the math scores was 83x13=1079 points.

Therefore Billy's score on the test was 1079-1020=59, answer choice (B).

Remember to work with the sums!

Real Problems that Use
ACT® Secret #10: Averages

page 164, problem 4

So far, the sum of the student's scores is: $65+73+81+82=301$. To earn an average of 80, the sum must become $80\cdot 5=400$. So the grade on the next test must be $400-301=99$.

(J)

page 167, problem 14

This is an easy calculation. Just add the numbers of Algebra I students and divide to get the average:

$$\frac{24+25+29}{3}=26$$

(H)

page 450, problem 1

So far, the sum of Kalino's tests is $85+95+93+80=353$. To earn an average of 90, the sum must become $90\cdot 5=450$. So the grade on the next test must be $450-353=97$.

(E)

page 452, problem 11

The total cost of ten notebooks is going to be $(2.50)\cdot(9)=22.50$. Therefore the average cost per notebook is $\frac{22.50}{10}=2.25$

(D)

page 465, problem 59

The sum for males is

$300\cdot 45=13,500$.

The sum for females is

$200\cdot 35=7,000$.

The average for the entire population is

$$\frac{13,500+7,000}{300+200}=41$$

(B)

ACT® Secret #11: Make a List

Sometimes a problem cannot be solved with algebra and it is structured so that backsolving won't help. In these cases, it is necessary to make a list of possible answers to find a solution. It is best to structure the list in the form of a table.

Typically, what you have to do is organize the list and then look for a pattern.

EXAMPLE PROBLEM:

If the sum of two distinct prime numbers is odd, the value of one of them must be:

(A) 0

(B) 1

(C) 2

(D) 3

(E) 4

SOLUTION:

Construct a table of prime number and their sums, then look for a pattern.

	2	3	5	7
2	----	5	7	9
3	5	----	8	10
5	7	8	----	12
7	9	10	12	----

Notice the cells of the table which are odd. The only odds appear in the row or column involving 2. The other cells are all even. Therefore the answer is (C).

GUESS: This is a great problem for guessing because 0, 1 and 4 are not prime numbers. If you knew this, you could have guessed either (C) or (D).

Real Problems that Use ACT® Secret #11: Make a List

page 175, problem 49

In order to see what is happening, make a list:

row	dots	cumulative total
1	1	1
2	3	4
3	5	9
4	7	16
5	9	25

The "trick" is to realize that what is required is the cumulative total, i.e., the number of dots in all of the first n rows. It is easy to see that that number is the square of the rows. (D)

page 313, problem 40

This is organized trial and error. Try squares of different numbers to determine the answer.

n	n^2
31	961
32	1,024
99	9,801
100	10,000

You can see that to obtain a perfect square with four digits, the square root must have two digits.

(G)

page 314, problem 44

Remember that two is the only even prime number. So we only have to check odd numbers:

n	prime?
31	Yes
33	No, 3
35	No, 5
37	Yes
39	No, 3
41	Yes
43	Yes
45	No, 5
47	Yes
49	No, 7

There are five prime numbers on the list.

(G)

ACT® Secret #12: Roman Numerals

Typically, an ACT® can be expected to have one or two Roman numeral problems. They are a variation of the multiple choice problem.

The way to approach a Roman numeral problem is to treat each Roman numeral like a true-false question. Mark each Roman numeral as true or false, then match the pattern of your answers to the answer choice.

EXAMPLE PROBLEM:

Two sides of a triangle are each 5 units long. Which of the following statements MUST BE true?

I. At least two of the angles of the triangle are congruent.

II. The third side of the triangle must be less than 10 units long.

III. The triangle is equilateral.

(A) I only

(B) II only

(C) III only

(D) I and II only

(E) I and III only

SOLUTION:

I. True. If two sides are congruent then the triangle is isosceles and its base angles are congruent.

II. True. The triangle inequality states that any side of a triangle must be less than the sum of the other two sides.

III. False. The triangle could be equilateral, but it might not be.

The answer is (D).

Real Problems that Use ACT® Secret #12: Roman Numerals

page 315, problem 49

The setup on this problem is not at all typical of a Roman Numeral problem. The key fact that they are testing is $|x| = \sqrt{x^2}$. This is something that sometimes is not taught in school.

In this problem, we need to simplify each expression:

I. $\sqrt{(-x)^2} = \sqrt{x^2} = |x|$

II. $|-x| = |x|$

III. $-|x|$ cannot be simplified

I and II are equal, but III is on its own.

(A)

page 461, problem 47

Take a close look at the diagram. It is clear that water was being pumped into the pool at a constant rate. Then the flow was slowed down, and the water level increased slightly.

I. False. The flow was slowed, not increased.

II. True. The flow was slowed.

III. True. If the drain were opened and the pump remained on, the water level could increase slightly if the pump is faster than the drain.

(E)

page 462, problem 48

I. True. The graph is a straight line segment, so its slope is constant. The value of the slope is

$$m = \frac{11-3}{4-0} = \frac{8}{4} = 2$$

II. True. The height of the graph ranges from 3 to 11.

III. False. A zero occurs when the graph crosses the x-axis. This line segment never crosses the x-axis.

(G)

Tracking Log for Practice ACT® Tests

Date	Test	Correct	Wrong	Raw Score*	Scaled Score**

*Answers begin on page 587 of The Real ACT Prep Guide. The sum of correct and wrong problems should be 60. The raw score is equal to the number of problems that were answered correctly.

**Find the scaled score from the grids at the end of The Real ACT Prep Guide. The first math grid is on page 592.

Notes

Breinigsville, PA USA
15 November 2009
227585BV00001B/35/P

9 780615 253992